Timothy Harley

Lunar Science

Ancient and modern

Timothy Harley

Lunar Science
Ancient and modern

ISBN/EAN: 9783337034528

Printed in Europe, USA, Canada, Australia, Japan

Cover: Foto ©berggeist007 / pixelio.de

More available books at **www.hansebooks.com**

LUNAR SCIENCE:

ANCIENT AND MODERN.

BY THE

REV. TIMOTHY HARLEY, F.R.A.S.,

AUTHOR OF "MOON LORE," ETC.

> " Heaven's ebon vault,
> Studded with stars unutterably bright,
> Through which the moon's unclouded grandeur rolls,
> Seems like a canopy which love has spread
> To curtain her sleeping world."
>
> *Shelley's "Queen Mab," iv.*

> " The man who has seen the rising moon break out of the clouds at midnight has
> been present like an archangel at the creation of light and of the world."
>
> *Emerson's " Essay on History."*

LONDON:

SWAN SONNENSCHEIN, LOWREY & CO.,

PATERNOSTER SQUARE.

1886.

Butler & Tanner,
The Selwood Printing Works,
Frome, and London.

NIGHT IN THE DESERT.

"How beautiful is night !
A dewy freshness fills the silent air ;
No mist obscures, nor cloud, nor speck, nor stain,
    Breaks the serene of heaven :
    In full-orbed glory yonder moon Divine
    Rolls through the dark blue depths :
        Beneath her steady ray
        The desert circle spreads,
Like the round ocean, girdled with the sky.
        How beautiful is night ! "

Southey's " Thalaba."

THE MIDNIGHT OCEAN.

" The mighty moon she sits above,
    Encircled with a zone of love,
    A zone of dim and tender light
    That makes her wakeful eye more bright :
    She seems to shine with a sunny ray,
    And the night looks like a mellow'd day !
    The gracious mistress of the main
    Hath now an undisturbèd reign,
    And from her silent throne looks down,
    As upon children of her own,
    On the waves that lend their gentle breast
    In gladness for her couch of rest ! "

Wilson's " Isle of Palms."

CONTENTS.

———◆◆◆———

LUNAR SCIENCE.

I.

INTRODUCTION.

" By Thy command, the moon, as daylight fades,
  Lifts her broad circle in the deep'ning shades ;
Array'd in glory, and enthroned in light,
  She breaks the solemn terrors of the night ;
Sweetly inconstant in her varying flame,
  She changes still, another, yet the same !
Now in decrease, by slow degrees she shrouds
  Her fading lustre in a veil of clouds ;
Now at increase, her gathering beams display
  A blaze of light, and give a paler day ;
Ten thousand stars adorn her glittering train,
  Fall when she falls, and rise with her again,
And o'er the deserts of the sky unfold
  Their burning spangles of sidereal gold :
Through the wide heavens she moves serenely bright,
  Queen of the gay attendants of the night." [1]

' Behold! a new spectacle of wonder! The moon is
making her entry on the eastern sky. See her rising in
clouded majesty ! opening, as it were, and asserting her

original commission to rule over the night : all grand and stately, but somewhat sullied in her aspect. However, she brightens as she advances, and grows clearer as she climbs higher, till at length her silver loses all its dross ; she unveils her peerless light, and becomes the beauty of heaven, the glory of the stars (Eccles. xliii. 9), delighting every eye, and cheering the whole world with the brightness of her appearance and the softness of her splendours." [2]

THIS passage from a once noted but now neglected book, eloquently directs us to the sphere of our present inquiries. Whatever our degree in the artificial scale of human society, whatever our profession in the arduous service of human toil, we all are more or less moved by " the divinity that stirs within us " to feel an interest in the edifices of the Divine Architect and the embellishments of the Divine Artist. Sir George Cornewall Lewis said truly: " The history of astronomy has numerous points of contact with the general history of mankind ; and it concerns questions which interest a wider class than professed astronomers, for whose benefit the existing histories have been mainly composed." [3]   No scripture, whether spiritual or scientific, which is of public interest, ought to be "of private interpretation." As an instinctively religious being, man is constitutionally concerned in the sublime science which, more perhaps than any other department of physics, has enabled him to look "through nature up to nature's God." If "an undevout astronomer is mad," the converse is at least measurably true : unastronomic piety is mad. For if science without devotion be defective, because the spiritual is the necessary inspiration of the material, devotion without science is also defective, because the material is the necessary embodiment of the spiritual. Science without religion is a corpse ; religion without science is a ghost. Immanuel Kant, in a splendid sentence oft quoted, says : "Two things fill the mind with ever new and increasing admiration and awe, the oftener and the more steadily we reflect on them—*the starry heavens above and the moral law within.*" ⁴ Let both be combined—God in creation and God in conscience : we shall then have described a circle, or two concentric circles, within which will be embraced the whole compass of human knowledge. Let both be followed—the light of stars, which Carlyle calls "street lamps of the city of God," and the light of conscience, which Butler calls "the candle of the Lord within us:" we shall then no longer "walk in darkness, but have the light of life." An eminent member of the Institute of France concludes a work on astronomy with some valuable words. After saying that "all truly great men have been believers in God," M. Rambosson adds : " It seems to me that all these considerations tend to demonstrate that the bond which in ancient times united astronomical science and religion has its origin in the very nature of man, and his necessary relations with the universe ; that the idea of causation leads up to the recognition of the Supreme Being as a rigorous and inevitable outcome of the laws of the mind ; and that the universe, being His natural expression, renders Him present to our sentiment." [5] Science and religion God hath joined together : he who would put them asunder is the friend of neither, but the enemy of both.

Of all the heavenly bodies none has a greater attraction upon our thought and attention than that beautiful satellite of our own earth, which from the earliest period has been to men a world of wonder and worship, speculation and study. It was the first object of Galileo's " incredible delight," for it was :—

"The moon, whose orb
Through optic glass the Tuscan artist views
At evening from the top of Fiesolé;
Or in Valdarno, to descry new lands,
Rivers, or mountains, in her spotty globe."

From that telescopic observation of "the starry Galileo," when, to use his own words, the moon appeared as near as if it had been distant only two semi-diameters of the earth, is dated the new era of modern selenography. For as Mr. Neison says, "Galileo on turning his telescope to the moon may be regarded as the first to substitute facts for conjecture with regard to the condition of the moon's surface."[6] Subsequent observers, through improved instruments, have unremittingly peered into its shining face, and to-day our acquaintance with the moon is so intimate that it is no longer a *terra incognita*, but a land of whose past history and present condition we expect to be fully informed at no very distant future. In the words of Sir John Herschel, "The physical constitution of the moon is better known to us than that of any other heavenly body. By the aid of telescopes we discern inequalities in its surface, which can be no other than mountains and valleys."[7]

As evidence of rapid progress in selenography, we may refer to those marvellous maps
of the moon which have been produced in
Germany, where science in every department
has made astonishing advance. Beer and
Mädler's chart of *Der Mond*, which may be
seen in the British Museum, measures thirty-
seven inches in diameter, and is said to indicate
the positions of 7,735 lunar craters.[8]   And Dr.
Schmidt's map, which as a whole is between
six and seven feet in diameter, is divided into
twenty-five sections, all photo-lithographed from
the original sheet. This chart marks the
positions of no less than 32,856 craters on the
moon.[9]   From the days of Galileo until the
present the telescope has revealed wonders ;
but now, in addition to micrometers and other
astronomical aids to the power of the human
eye, we have the assistance of lunar photography, which, in such able hands as those of
Mr. De la Rue of London, and Mr. Ruther-
furd of New York, will furnish us with portraits
of the moon, increasingly exquisite, and with
lunar maps far more accurate than are any of
our maps of the earth. Mr. De la Rue has
already produced a fine photograph of the

moon, 38 inches in diameter; and the "Moon Committee" of the British Association has taken up the work in earnest. Such workers are worthy of high honour. Astronomers rank among the ordained ministers and official interpreters of nature : and their labours merit the admiration of all who believe in the Bible of Creation, which reveals not only unvarying law written in letters of unsullied light, but also lessons of goodness in greatness, which through the expositions of science are rendered intelligible to all.

II.

THE MOON'S DISTANCE.

"Daughter of heaven, fair art thou! the silence of thy face is pleasant! Thou comest forth in loveliness. The stars attend thy blue course in the east. The clouds rejoice in thy presence, O moon! They brighten their dark-brown sides. Who is like thee in heaven, light of the silent night? The stars are ashamed in thy presence They turn away their sparkling eyes. Whither dost thou retire from thy course, when the darkness of thy countenance grows? Hast thou thy hall, like Ossian? Dwellest thou in the shadow of grief? Have thy sisters fallen from heaven? Are they who rejoice with thee, at night, no more? Yes! they have fallen, fair light! and thou dost often retire to mourn. But thou thyself shalt fail one night, and leave thy blue path in heaven. The stars will then lift their heads: they, who were ashamed in thy presence, will rejoice. Thou art now clothed with thy brightness. Look from thy gates in the sky. Burst the cloud, O wind! that the daughter of night may look forth, that the shaggy mountains may brighten, and the ocean roll its white waves in light." [10]

SUPPOSE that we set out upon an imaginary journey to the moon. We shall have to pass through a cold medium; for the learned have calculated, upon certain data, that the temperature of space is about ninety degrees below zero. We are refrigerated till we shiver at the thought. Our goal is a long way off; nearly two hundred and forty thousand miles. This is in round numbers, which seem the right numbers in speaking of the spheres. Messrs. Nasmyth and Carpenter assign 238,790 miles from the earth as the moon's mean position; while Mr. Norman Lockyer, still more precise, says that the moon revolves "at an average distance of only 238,793 miles, which is equal to about ten times round our planet." " This measurement is done by determining the moon's parallax by means of trigonometry. But Sir G. B. Airy mentions another method, which, if not so exact in its results, is more easily understood and verified. He says, " There is a phenomenon of the moon observed frequently, in the interpretation of which there can be no mistake, namely, eclipses of the moon. We see that the moon, in her motions through the stars, dips into something which obscures her. There cannot be a doubt that it is the shadow of the earth. The moon goes into this shadow on one side, and comes out of it on the other side. The time which the moon occupies in passing through this shadow is, roughly speaking, four hours. The moon, then, is at such a distance that in passing through the shadow of an object as big as the earth, she is occupied only four hours. The moon, therefore, in her course describes the breadth of the earth in four hours; in one day she describes six times the breadth; and as thirty days is a rough measure for the time of her revolution, she describes in one revolution 180 times the breadth of the earth, and therefore the whole circumference of the moon's orbit is something about 180 times the breadth of the earth, and the diameter of the moon's orbit is about 60 times the breadth of the earth. Therefore the moon is distant from us by about thirty times the earth's breadth." [12]   The equatorial diameter of the earth we know is 41,848,380 feet, or just under 7,926 miles, which multiplied by 30 gives the result which is not so very far from right. But the orbit of the moon being an ellipse, and not a circle, this distance varies with her motions. During the *perigee*, when, as the word means, she is nearer to the earth, and appears larger, she is

then within 225,719 miles; but when in *apogee*,
or away from the earth, and seemingly smaller,
she is then 251,947 miles distant : a difference
of 26,228 miles.   Two hundred and forty
thousand miles is a long journey, but it is
wonderfully diminished by comparison with the
space which separates us from our companion
planets, to say nothing of the distances of the
fixed stars.   Venus, our nearest sister planet,
is never nearer than 24,000,000 miles, or a
hundred times the distance of the moon.   As
Dr. Leitch says, " While Neptune is a mile
distant, the moon is, on the same scale, only
about six inches.   And man, even when he
could form no idea of the real distance, ever
looked to the moon with a familiarity which he
could feel towards no other heavenly body." [13]
If a viaduct or suspension bridge could be built
across the intermediate ocean of space, we
might walk to the moon ; but at the rate of
four miles an hour, we should require but a
little less than seven years.   With a good
horse, trotting two hundred and forty miles a
day, we could get to our destination within
three years ; while with a railway, and such a
train as the " Flying Dutchman," doing sixty

miles an hour, we should gain the terminus
within six months. No time is allowed in this
reckoning for refreshment : but, of course, on
such a heavenly journey, we should be above
all earthly things. The astronomical Arago
has a much quicker method of locomotion than
any which we have suggested. He says : " It
is known that a 24 pounder traverses, at the
utmost, 1,312 feet per second, at the instant of
quitting the cannon's mouth. This velocity is
equivalent to $2\frac{1}{2}$ miles in ten seconds, to 15
miles in a minute, to 900 miles in an hour, to
21,600 miles in a day. Hence a cannon ball
would pass from the earth to the moon in
eleven days." [14] That would be " shooting the
moon " with a vengeance! Indubitably, it
would be quick travelling; and if we were
rolled up in that sort of conveyance we should
arrive in haste, and take the immortal man in
the moon by surprise. Such express speed
would even surpass that American train which,
according to one of the latest transatlantic
stories, was rushing along so rapidly, that a
passenger who thrust his head out of the
window to salute a female friend at one station,
kissed an old coloured woman at the next
station instead.

While we are making for the moon, we may compare notes with the ancients. Eratosthenes, the Alexandrian astronomer, who died B.C. 194, and who must be ever memorable as the first measurer of the earth's magnitude, said that the moon was distant from our globe 78,000 stadia ten times multiplied. The puzzle is to ascertain the precise length of a stadium. We can attain to only an approximation of its value by the following method. Eratosthenes, whose observations were made by means of the shadow of a gnomon, measured an arc of the terrestrial meridian, whose result gave the circumference of a great circle as equal to 250,000 stadia in length. Now the earth's circumference is known to be about 25,000 miles, whence we conclude that a stadium is the tenth of a mile. According to this calculation, Eratosthenes believed the moon to be 78,000 miles from the earth : a little less than a third of its real distance. The Brahmins of India erred in another direction, for they said that the sun was nearer to the earth than the moon, probably because the sun was known to be hotter and brighter. To this day the Puránas of India teach that " the moon is

twice as far from the earth as the sun." [15]
Flammarion cites an old Italian system of astronomy which gave the moon's distance "from
the centre of the earth to the inner side of the
heaven of the moon" as 107,936 miles. This
author also states that to the Egyptians the
sun was only 369 and the moon 246 miles
away. [16] Last century Dr. Rogers, writing of
the moon, estimated that " her distance from
the earth must be 216,004 miles, which is
23,996 miles less than the general estimation."[17]
Enough has now been said to show that in
respect of the moon's distance we "understand
more than the ancients "; that correct calculation of celestial intervals is an accomplishment
of that " modern thought" which some modern
talkers, *without thought*, idly abuse; and that
we are doubtless orthometrical within a very
little when we give the mean distance between the centres of the earth and moon as
238,793 English miles. Telescopes with a
magnifying power of 1,000 will bring the moon
within 240 miles of our observation ; while a
power of 2,400 times, if it were practicable,
would enable us to see the moon from the earth
as we may see Mont Blanc from Lyons, a

distance of 100 miles. Evidently man has an affinity with the infinite.

"I was yesterday about sunset walking in the open fields, till the night insensibly fell upon me. As I was surveying the moon walking in her brightness and taking her progress among the constellations, a thought rose in me which I believe very often perplexes and disturbs men of serious and contemplative natures. David himself fell into it in that reflection. 'When I consider the heavens the work of Thy fingers, the moon and the stars which Thou hast ordained, what is man that Thou art mindful of him, and the son of man that Thou regardest him!' In the same manner, when I considered that infinite host of stars, or, to speak more philosophically, of suns, which were then shining upon me, with those innumerable sets of planets or worlds, which were moving round their respective suns; when I still enlarged the idea, and supposed another heaven of suns and worlds rising still above this which we discovered, and these still enlightened by a superior firmament of luminaries, which are planted at so great a distance that they may appear to the inhabitants of the former as the

stars do to us; in short, whilst I pursued this
thought, I could not but reflect on that little in-
significant figure which I myself bore amidst
the immensity of God's works." [18]

III.

THE MOON'S SIZE.

"The moon is up, and yet it is not night—
  Sunset divides the sky with her—a sea
Of glory streams along the Alpine height
  Of blue Friuli's mountains ; heaven is free
  From clouds, but of all colours seems to be,—
Melted to one vast Iris of the West,—
  Where the day joins the past Eternity ;
  While, on the other hand, meek Dian's crest
Floats through the azure air—an island of the blest !" [1]

HAVING passed "from the earth to the moon,"
we might follow M. Jules Verne in his sequel,
and journey "around the moon," to take her
dimensions. It would be cruel to torture the
weary minds of readers with many figures,
therefore it may be sufficient to state that the
moon's diameter, which is the length of a hypo-
thetical line drawn through its rotund body,
is, according to Flammarion, 2,153 miles, and,
according to Nasmyth and Carpenter, 2,160.
The former gives the total surface as 14,568,000

square miles, and the latter give its area as
14,657,000 square miles, one-half of which, or
7,328,500 miles, is the area of the hemisphere
presented to our earth.   Further, to show that
this work of lunar mensuration has been done
thoroughly, we are told that the volume of the
moon contains 5,276,000,000 of cubic miles of
solid matter.   Still, the moon's diameter is less
than a third of that of our own great globe, and
its bulk scarcely more than a forty-ninth part
of that of the earth.   As a knowing German
puts it, " The surface of the earth is about
fourteen times larger than that of the moon,
and, in solid contents, is fifty times greater.
To an observer in the moon the earth must
appear 13·5 times larger than the moon appears
to us." [20]   The late president of the Royal
Astronomical Society tells us that " the mass of
the moon is about the $\frac{1}{81}$ part of that of the
earth, or, it would require eighty-one moons to
make a globe of corresponding weight to that
of the world in which we live." [21]   It may
interest some to learn that the earth's weight
is 6,069,000,000,000,000,000,000 tons !   This
excludes the air, which weighs less than a
millionth part of the earth.

Of course, compared with the sun, the moon is a very small taper indeed : the diameter of the day-star being over four hundred times that of the night-light. This may be inferred from the fact that they appear equally large, though the distance of the sun is about four hundred times that of the moon. In fact, it would require a million and a half moons to equal in magnitude one sun. An astronomical author informs us that "if the whole sky were covered with full moons they would scarcely make daylight, for Bouger's experiments give the brilliancy of the full moon as only $\frac{1}{300,000}$ that of the sun." [22] Only fancy 300,000 full moons ! A hundred times more moons than there are stars visible to the naked eye ! What an illumination ! The heavens could not contain them. Yet their united lustre would barely equal the resplendent sovereign of the system ! An American astronomer augments the disparity by a novel experiment. "Professor G. P. Bond compared the light of the moon with that of the sun by placing in the sun's light a glass globe with a silvered surface, and comparing the brightness of the reflected image of the sun with an artificial light, and after-

wards comparing the light of the full moon
with the same standard. He hence inferred
that the light of the sun was 470,000 times
that of the full moon."[23] And another
American astronomer, Professor Newcomb, of
Washington, says that "the most careful determination yet made is by Zöllner, who finds the
sun to give 619,000 times as much light as the
full moon. This result is probably quite near
the truth."[24] We are dazzled and dazed with
the effulgence of such wondrous worlds. And
when we further consider the moon, with its
diameter of over 2,000 miles, and its circumference of 6,000 miles, travelling round the earth
at the rate of 2,000 miles an hour, while at the
same time it is carried forward with us round
the sun in our vast orbit of 600,000,000 of miles
at the rate of 68,000 miles an hour, our wonder glows with worship, and admiring science
becomes an adoring psalm.

It will be interesting to inquire what the
ancients knew of the moon's size. Thales,
the founder of the Ionic school of philosophy,
B.C. 600, made some attempt to measure the
apparent magnitude of the moon. Anaximander, his disciple, said it was a circle nine-

teen times bigger than the earth; Parmenides, that it was equal in brightness to the sun. Aristarchus of Samos, a diligent and philosophical astronomer, endeavoured to accurately estimate the magnitudes of the sun and moon. He flourished about B.C. 276, and "made the diameter of the moon to that of the earth greater than 43 to 108, and less than 19 to 60 : this last determination is not far from the truth." [25] Other Greeks made other guesses. " One of them said that the moon was as large as that part of Greece once known as the Peloponnesus, but now called the Morea, and was laughed at for his boldness." [26] Brave old Greek ! and he was no other than Anaxagoras of Clazomenœ, the master at whose feet sat Socrates, and the man to whom science and philosophy were of higher worth than wealth, or honour, or life. He was accused of impiety at Athens, and though defended by Pericles, was condemned to death. When asked whether his body should be carried back to his own country, he answered, " No, for the road which leads to the other side of the grave is as long from one place as another ! " As a recent author says, " Anaxagoras is interesting as being the first martyr of science. He was accused of impiety at Athens for teaching that the moon, then regarded, with the other heavenly bodies, as divine, was of the same nature as the earth, traversed by hills and valleys, and probably inhabited." [27] Twenty-three centuries have elapsed since Anaxagoras suffered, and we have not heard of the last martyr of science yet! But to return to the moon's dimensions. The Stoics held that it was bigger than the earth, and the Peripatetics, with all their "walking about," arrived no nearer to the truth. Lucretius, among the Romans, held that the moon was " of no larger a dimension than she appears to our eyes as we observe her," which opinion displays a singular ignorance or forgetfulness of the rules of perspective, an art not unknown to the ancients. The Hebrews were a religious people, and not scientific. We shall therefore look in vain for any effort on their part to work out the problems of nature. Their allusions to the hosts of heaven are numerous and often highly poetical, but not at all philosophical. For example, on the point in question : " The Talmud states that when the sun and moon were first created they were of equal size. The moon became jealous of the sun, and she was reduced in bulk. The moon then appealed to God, and she was consoled by the promise that Jacob, Samuel, and David were to be likewise small. As, however, some injustice seemed to have been committed, God ordained 'a sin-offering' on every new moon, because the moon had become less than the sun." [28] The Buddhist doctrine is that "the disc of the moon is 40 yojanas in diameter, and 147 in circumference." [29] But what is a yojana? is another philological knot. Mr. Hardy tells us that "the length of the yojana is a disputed point. By the Singhalese it is regarded as about sixteen miles in length, but by the Hindus of the continent as much shorter." [30] From Horace Wilson, the Sanskrit scholar, we learn that a *yojana* is "a measure of distance equal to four kro'sas, which at 8,000 cubits, or 4,000 yards to the *kro'sa* or *ko's*, will be exactly nine miles; other computations make the yo'jana but about five miles, or even no more than four miles and a half." [31] We thus see that the diameter of the Buddhist moon does not exceed 360 miles. What would the lunarians say to these few figures? Especially if they knew what a French author says of them in relation to the smallness of their world. "This would probably not prevent its inhabitants (if there are any) from fancying themselves superior to us, and believing us to be their servants rather than their masters; for it is generally known that the smaller people are, the more vanity they possess."[32] The *modus operandi*, in determining the diameter and mass of the moon, is detailed by Dr. Lardner in his *Handbook of Astronomy* (sections 2640–44), and described with greater brevity by Mrs. Somerville, who says, "The mass of the moon is determined from several sources—from her action on the terrestrial equator, which occasions the nutation in the axis of rotation; from her horizontal parallax; from an inequality she produces in the sun's longitude; and from her action on the tides. The apparent diameters of the sun, moon, and planets are determined by measurement; therefore their real diameters may be compared with that of the earth; for the real diameter of a planet is to the real diameter of the earth, or 7,899 miles, as the apparent diameter of the planet to the apparent diameter of the earth as seen from the planet, that is, to twice the parallax of the planet." [33]   In briefest form, we may say that the ratio of the moon's mass to the masses of the earth and the sun is computed by the observed effects of lunar attraction ; while the diameter is ascertained by measurement based upon the lunar parallax.

IV.

THE MOON'S SHAPE.

" Meanwhile the moon,
Full orb'd and breaking through the scatter'd clouds,
Shows her broad visage in the crimsoned east.
Turned to the sun direct, her spotted disk—
Where mountains rise, umbrageous dales descend,
And caverns deep, as optic tube descries,
A smaller earth—gives all his blaze again,
Void of its flame, and sheds a softer day.
Now through the passing cloud she seems to stoop,
Now up the pure cerulean rides sublime.
Wide the pale deluge floats, and streaming mild
O'er the skied mountain to the shadowy vale,
While rocks and floods reflect the quivering gleam,
The whole air whitens with a boundless tide
Of silver radiance, trembling round the world." [34]

ALIGHTING in the lunar world, we find a body,
neither round, square, nor exactly oval, but of
that lovely figure called an ellipsoid, or oval-like.
The earth, we know, is an oblate spheroid,
which the dictionaries define as a sphere de-
pressed at the poles; somewhat resembling a

candidate for parliamentary honours, when he finds himself defeated. Columbus imagined that the earth had the shape of a pear. The philosophizing Knickerbocker preferred an orange; for that "small, brisk-looking old gentleman" begins his description of the world as follows: "According to the best authorities, the world in which we dwell is a huge opaque, reflecting, inanimate mass, floating in the vast ethereal ocean of infinite space. It has the form of an orange, being an oblate spheroid, curiously flattened at opposite parts for the insertion of two imaginary poles, which are supposed to penetrate and unite at the centre; thus forming an axis on which the mighty orange turns with a regular diurnal revolution." [35] The moon also has a conformity to the *golden* fruit; for although *orange* is probably Persian, we enjoy the etymology which connects it with the Latin for gold. "Small photographs of the full moon look so much like photographs of a peeled orange, that, as Wendell Holmes notes, many persons suppose astronomers have substituted the orange for the moon, so as to save themselves trouble." [36] This is a fruitful suggestion; and if lecturers on lunar physics, especially to juvenile audiences, were to pass round oranges as illustrations of the moon's shape, we are confident that their prelections would be more delectable and far less dry. Heraclitus, "the obscure philosopher" of Ephesus, compared the moon to a boat; others to a circle. The former meant the *new* moon; the latter, Luna in *full* dress. Empedocles of Sicily said that it resembled a basin or platter. This reminds us that the Indians of North America call the earth the big plate where all the spirits eat, and they think that it enlarges with the increase of vegetation and inhabitants. Some of the Winnebagoes believe that the earth is oval on the top, and flat at the bottom. Herodotus ridiculed the idea of the earth's circularity. He says, "Many even now commit the ludicrous and ignorant error of drawing a map of the earth, in which it is represented of a circular form, as if its outline were traced with a compass, and the ocean is made to flow round it." [37] Berosus of Babylon thought the moon a globe with one side luminous, and the other of a sky-blue. To the Stoics of Greece our satellite was round as a sphere: while her circular shape suggests to the Hindus the idea of a ring or pearl; and the pearl-moon, from its ambrosial humour, has a fine water.

V.

THE MOON'S SUBSTANCE.

" The moon's the earth's enamour'd bride ;
   True to him in her very changes,
   To other stars she never ranges :
     Though crossed by him, sometimes she dips
Her light, in short offended pride,
   And faints to an eclipse.

The fairies revel by her sheen ;
   'Tis only when the moon's above
   The fire-fly kindles into love,
     And flashes light to show it :
The nightingale salutes her queen
   Of heaven, her heav'nly poet." [38]

VARIOUS opinions, ancient and modern, have been held of the moon's substances. Some regarded it as a rotund rock : a kind of hard head, without trunk or limbs, whose corrugated brow bore witness to its old age and rough experience. Others have had no doubt that it was of much lighter material than the earth ; that is, of less mean density. Some among the ancients thought that the moon was a celestial kind of earth, where our mud earth was pitied. Others that it was a sort of pumice-stone. Socrates was credited with saying, "that the sun is a stone, and the moon earth." [39] Anaximander taught that the moon was full of fire; Anaximenes, that it was the same as the earth ; Empedocles, that it was congealed mist hardened by fire ; Xenophanes, the father of the Eleatics, that it was a thick, compact, felted cloud ; Parmenides, that it was a mixture of fire and air ; Heraclitus, that it was of earth overspread with mists ; Anaxagoras and Democritus, that it was a "solid and firm body, all fiery, containing champaign grounds, with mountains and valleys ;" Plato, that it was a fiery substance ; Pythagoras, that its body was of the nature of fire ; the Stoics, that it was a mixture of fire and air. The Buddhists say that " within it is composed of crystal, and its surface is of silver ; so that both its surface and inner material are extremely cold." [40] Hegel held it to be a "material crystallization." Yet with this mass of wisdom before them, certain scientific men affirm that the moon's physical history is a physical mystery, and that its substance is substantially unknown. With due deference to scientific men we profess to be better informed. We are in possession of cumulative evidence that the moon is made of green cheese. We proudly refer to such re-doubtable authorities as Samuel Butler and François Rabelais, with many more. The rising generation seldom make mistakes, and one of their nursery rhymes is of three children who went hunting by night :—

> " One said it was the moon,
> Another said, Nay;
> A third said it was a cheese,
> And half o't cut away."

The heroic *Hudibras* acknowledged the prevalency of this opinion, for

> " He made an instrument to know
> If the moon shine at full or no;
> That would, as soon as e'er she shone, straight,
> Whether 'twere day or night demonstrate;
> Tell what her di'meter to an inch is,
> And prove that she's not made of green cheese." [41]

Rabelais records that Gargantua " thought the moon was made of green cheese." [42] Further, in that " progeny of learning," as Mrs. Malaprop would have called it, the library of the British Museum, we discovered the follow-

ing passage, three centuries old. Its spelling
is beautiful to behold; its poetry most exquisite.

" Such is the fashyon of the worlde now a dayes,
  That the symple innosaintes ar deluded
  And an hundred thousand divers wayes
  By suttle and craftye meanes shamefullie abused.
  And by strength, force, and violence oft times compelled
  To believe and saye the moune is made of a grene chese
  Or ells have great harm, and parcace their life lese." [43]

How those who deride modern thought will
glory in such ancient wisdom as this! Our
astute ancestors indulged in no senseless jar-
gon about nebular hypotheses, laws of gravitation, undulatory theories, centrifugal and
centripetal forces; they knew that the earth
was stationary and flat, that all of the heavenly
armies marched over their heads, every twenty-
four hours, for men to review them; and that
the moon was a single Gloucester cheese, and
its inhabitants mites. This, probably, is the
origin of that wicked witticism, "When is a
cheese like the moon?" Answer, "When it is
high and *mity.*" The French, in allusion to the
cheesemonger's moon, say, "Il veut prendre
la lune avec les dents" (*He wants to take the
moon between his teeth*). It must have been

with kindred feeling that the Wiltshire moon-raker, who saw the moon reflected in a pond, and took it to be a cheese, tried to fish it out with a rake. This may appear an idle notion, but beyond question it is widespread and of considerable antiquity. In the Highland fable of the wolf and the fox, the wolf lost his tail through supposing the moon was a cheese. The fox said, " I smell a very nice cheese, and (pointing to the moonshine on the ice) there it is, too." " And how will you get it ? " said the wolf. " Well, stop you here till I see if the farmer is asleep ; and if you keep your tail on it, nobody will see you or know that it is there. Keep it steady. I may be some time coming back ! " So the wolf lay down, and laid his tail on the moonshine in the ice, and kept it for an hour till it was fast. The fox then gave the alarm, the farmer and his wife came with sticks to kill the wolf, who, not wishing to be killed, ran off, leaving his long tail behind him ; and that is why the wolf's tail is stumpy to this day." This fox and wolf story recalls the old Norse tale of " Why the bear is stumpy-tailed." The fox told the bear to cut a hole in the ice, and hold his tail therein for a fishing line.

When it began to smart he would know that the fish were biting. He did so, and the fox's *caudal* lecture cost the bear the extreme loss which all bears inherit. Strange to say, the above Highland legend is found, in another form, in the Talmud. " Rabbi Meir was a great allegorist ; it is said that he knew three hundred allegories relating to the fox alone. Of these but three fragments remain to us." One is as follows : " A fox said to a bear, ' Come with me, my good friend ; let us not quarrel : I will lead thee to another place where we shall surely find food.' The fox then led the bear to a fountain, where two buckets were fastened together by a rope, like balances. It was night, and the fox pointed to the moon reflected in the water, saying, ' Here is a fine cheese ; let us descend and partake of it with an appetite.' The fox entered his pail first, but being too light to balance the weight of the bear, he took with him a stone. As soon as the bear had gotten into the other pail, however, the fox threw this stone away, and consequently he rose, while the bear descended to the bottom." [45] That the moon is a cream cheese may be merely a *reductio ad absurdum*

to prove the futility of former attempts to
elucidate the enigmas of nature, one of which
was the constitution of distant worlds. The
material of the moon is a mystery not even
now altogether revealed. Most likely the
elements are much the same as those of the
earth. The main difference is doubtless one
of proportion in the composition rather than of
essence in the ingredients. For if we accept
the hypothesis of Laplace, that satellites are
masses thrown off from their primaries, as
those primaries had been previously thrown off
from the sun, we must infer that the constituent
materials of one globe are the same as those of
all the others having a common genesis. The
chemistry of creation is exceedingly simple in
its elementary substances, but infinitely com-
plex in its countless combinations. Analysis
has given us an insight into the synthesis of
our own globe, and "how infinitely is the
knowledge increased in interest, when we consider the probability of such being the materials
of the whole of the bodies of space, and the
laws under which these everywhere combine,
subject only to local and accidental varia-
tions!" [46] The chemistry of the future will

do more than demonstrate that creation is a *universe* "turned into one," or "combined in one whole " : it will show how totals grow from units, whose original UNUS is GOD.

VI.

THE MOON'S FORMATION.

" Now came still evening on, and twilight grey
  Had in her sober livery all things clad ;
  Silence accompanied ; for beast and bird,
  They to their grassy couch, these to their nests,
  Were slunk, all but the wakeful nightingale ;
  She all night long her amorous descant sung ;
  Silence was pleased : now glow'd the firmament
  With living sapphires : Hesperus, that led
  The starry host, rode brightest, till the moon,
  Rising in clouded majesty, at length
  Apparent queen, unveiled her peerless light,
  And o'er the dark her silver mantle threw." [47]

" IN beginning, God created the heavens and
the earth." This is true, not because it is
found in the Bible ; but it is found in the Bible,
because it is true. And it will be admitted
as axiomatic by all, save a few unreasonable
minds who admit nothing except their own
right to deny everything. This cosmogony in
a clause contains the five points of *a posteriori*
reasoning upon the genesis of the universe.

Firstly, the reasoner, commencing with himself, the *ego*, as the metaphysicians have it, sees that he is a part of a whole inhabited earth. *Secondly*, he observes that the earth is to him the centre of limitless heavens ; the grand totality in which his own world is but a unit. *Thirdly*, he discovers in all worlds what Sir John Herschel called the "appearance of manufactured articles " : they were created or made. *Fourthly*, he concludes that what was made must have had a maker, whom he calls God, or in Indo-Germanic language, the Brilliant Being, who spreads the light. *Finally*, he rests his weary wing on the boundary of *nature*, which means what is *born*; and in the moment of that birth, his *terminus a quo*, he finds the beginning of all things which exist. That such beginning took place but six thousand years since, he cannot believe with the sunlight of modern science shining upon his studies. Still, he does not ridicule those who in ancient times held otherwise ; he ridicules only those who at the present day claim for the ancients a scientific accuracy which they never possessed, and which they never claimed for themselves. He is, however, in entire har-

mony with the writer who affirms that "God
made two great lights ; the greater light to rule
the day, and the lesser light to rule the night."
How the sun and moon were made, that writer
neither knew nor pretended to know ; and the
how is a matter of much conjecture to the
present moment.   Let us unanimously and
reverently acknowledge the Divine Authorship
of nature, postulating a First Cause as a logical
necessity ; and then proceed to questions of
process and development, or evolution, without
fear or reproach.

Now, to turn to the formation of the moon.
Many astronomers, Nasmyth and Carpenter
among the number, maintain its origin to have
resulted through the diffusion of primordial
matter, whose particles, under the combined
action of impulse and gravitation, generated
cosmical heat.   For motion, as Locke pointed
out, manifests itself as heat ; and heat, as Professor Tyndall shows, is a mode of motion.
Particles of matter colliding by mutual gravitation would generate sufficient heat to reduce
the whole to a molten mass.   But with the ag-
gregation of matter, the generation of dynamical
heat would be diminished ; the material being

meanwhile condensed into a spherical planetary
body. Cooling would occur from radiation, first
upon the surface ; and consequently " with the
solidification of this external crust began the
year one of selenological history." [48] Through
cooling, the superficial matter of the moon
might contract into wrinkles like a shrivelled
apple or a rugous hand; and thus the de-
pressions may have occurred. Mr. Proctor
seems to think that there was a time when the
earth's vapour globe, extending to and envelop-
ing the moon, shrunk in its dimensions, leaving
the moon, a vapour nucleus, out in the cold ;
and that therefore, being smaller, it became
solid sooner than the earth. He does well to
add, " It is manifest that we have in the moon
a subject of research which has been by no
means exhausted." [49] The orientals, however,
think otherwise. The Hindus make an oblation of clarified butter, with the following
prayer : " Gods ! produce that [moon] which
has no foe, which is the son of the solar orb,
and became the offspring of space, for the
benefit of this world ; produce it for the ad-
vancement of knowledge, for protection from
danger, for vast supremacy, for empire, and for

the sake of Indra's organs of sense.   May
this oblation to the lunar planet be efficacious." [50]
The Eastern mind lingers in that poetic state
of adolescence which sees scientific objects in
a "dim religious light"; while the Western
mind is chafed by those limitations which
retard investigation.   But "with time and
patience, the mulberry-leaf becomes satin."

> "We have but faith ; we cannot know :
>   For knowledge is of things we see ;
>   And yet we trust it comes from Thee,
> A beam in darkness : let it grow."

The Mexican myth of the moon's formation
is that Tezcociztecal, following the example of
the hero who had become the sun, threw himself into a great fire ; "but the flames being
somewhat less fierce, he turned out less bright,
and was transformed into the moon." [51]   The
Esquimaux tradition is radically the same ;
and we may read it as science in a fable, or as
a fable in science.   "There was a girl at a
party, and some one told his love for her by
shaking her shoulders, after the manner of the
country.   She could not see who it·was in the
dark hut, so she smeared her hands with soot,
and when he came back she blackened his

cheek with her hand. When a light was brought she saw that it was her brother, and fled. He ran after her, followed her, and as she came to the end of the earth, he sprang out into the sky. Then she became the sun, and he the moon; and this is why the moon is always chasing the sun through the heavens, and why the moon is sometimes dark as he turns his blackened cheek towards the earth." [52] This masculine moon is of high antiquity.

VII.

THE MOON'S CONDITION.

" Queen and huntress, chaste and fair,
    Now the sun is laid to sleep,
Seated in thy silver chair,
    State in wonted manner keep.
      Hesperus entreats thy light,
      Goddess excellently bright !

Earth, let not thy envious shade
    Dare itself to interpose ;
Cynthia's shining orb was made
    Heaven to clear, when day did close.
      Bless us then with wishèd sight,
      Goddess excellently bright !

Lay thy bow of pearl apart,
    And thy crystal-shining quiver :
Give unto the flying hart
    Space to breathe how short soever ;
      Thou that mak'st a day of night,
      Goddess excellently bright !" [53]

O'KEW, the physiophilosopher of Zürich, who
taught that the moons have come, "not
mechanically, but dynamically, by polarization
according to eternal laws, according to the laws

of light," tells us in laconic language, "the moon is in itself dead." [54] We have met with other decisions just as peremptory as this, and have felt tempted to write *cadit quæstio*, and to timidly confess that the case allowed of no further inquiry. Especially after one of our own favourite poets had asked so pitifully of the moon :—

> " Art thou pale for weariness
> Of climbing heaven, and gazing on the earth,
>   Wandering companionless
> Among the stars that have a different birth,—
> And ever changing, like a joyless eye
> That finds no object worth its constancy?" [55]

Surely the moonbeams must be none other than the sheeted ghosts which haunt the sepulchres where lie the petrified remains of primeval lunarian life. But upon recovery of ourselves, we object to the argument being foreclosed in this fashion. It is not a century ago since Sir William Herschel wrote (April, 1787), " I perceive three volcanoes in different places of the dark part of the new moon. Two of them are already nearly extinct, or otherwise in a state of going to break out; the third shows an eruption of fire or luminous matter." No doubt the great astronomer was mistaken, and that what he saw was only "earth-light reflected from those parts of the moon's surface which have the least reflective capacity," But if there be no active volcanoes in the moon just now belching out their fires, we know that the lunar craters are so numerous that Galileo compared them to the eyes in a peacock's tail for multitude, and so immense that, compared with them, our Vesuvius and Etna are very small. Who can tell that some night we may not see one of these tremendous furnaces bursting into flame, and convincing amazed beholders that there is life in the old moon yet? Mr. Norman Lockyer, in his lucid *Lessons in Astronomy*, notes one possible exception to the rule that these volcanoes are extinct; and the end of our knowledge being the beginning of our ignorance, we wait for further light. A more material question, perhaps, is that of water and atmosphere upon the lunar surface. Kepler, Hevelius, and Ricciolus thought the dark portions were *existing* seas; but if the smooth plains on the moon be the dry bottoms of *former* seas, what has become of the water? Flammarion says, "Since the distant period of its formation in a fluid state, it has lost all its liquids and vapours, and now a linnet would die of thirst in the midst of the seas of the moon. These seas do not contain a drop of water. These, it will be said, are singular seas." [56]   Mr. Mallet, our leading seizmologist, lays down the rule that "without water there can be no eruption." If this be so, and if eruptions have taken place and are possible still, we press the question, What has become of the water ?   William Whiston said that a comet swept it away; as another comet produced a deluge on the earth, and replenished our seas. This author's account of Noah's flood is worth reproduction : "When the earth passed clear through the atmosphere and tail of the comet, in which it would remain for about two hours (as from the velocity of the earth, and the crassitude of the said tail on calculation does appear), it must acquire a large cylindrical column of vapours." [57]   And this sweep of the comet's tail was the source of the forty days of rain ! Well, Whiston was a very learned man ; but the "crassitude" of our minds prevents us from entertaining his theories with regard to the appearance of much water on the earth, and the *dis*-appearance of all water on the moon. Another authority says that the lunar fluids are all frozen solid. But if during the long lunar day the temperature of the moon's face under the burning sun rises to more than 500° Fahrenheit, much of the ice ought to melt, and be visible from the earth in masses of mist. A third theory is that the moon has opened its mouth and swallowed all the water into its capacious interior. This hypothesis is favoured by Guillemin, who thinks that the cavernous structure of the moon's interior would provide a receptacle for its ocean, from the depths of which the sun would be unable to dislodge more than traces of its vapour. [58] And a fourth opinion is that it is withdrawn to the other side of the moon. Even the illustrious Hansen held that the hidden half might possess water. Such an explanation is admissible and admirable only on the principle of Tacitus, *Omne ignotum pro magnifico est ;* which may be freely rendered, What we do not know supplies a magnificent substitute for what we would like to know. But the moon's centre of gravity is more than thirty miles farther away from the middle point of that diameter which is directed towards the earth. Therefore the moon is heavier on one side than she is on the other, like a loaded die ; and it is thought that the ocean must flow over to the heavier side. Certainly the farther side of the moon could accommodate the water if it be, in Sir John Herschel's words, "like a great lake basin, nearly forty miles deep." But now we are told that Gussew of Wilna has come to the conclusion that the moon is ellipsoidal, or egg-shaped, the egg being nearly round, and its smaller end and major axis being directed towards the earth. The probability, therefore, is that the other lunar hemisphere is much the same as the side which we see, with as much water and no more. To the existence of a lunar atmosphere, appearances and opinions founded upon them are unfavourable. Though if there be eruptions, we might not only infer the presence of water, but also expect that carbonic acid gas would form an envelope, albeit of extreme rarity. Both water and atmosphere must have been on the moon at one period, if our satellite has passed to its present deforma-tion through ages of life and beauty such as, happily for us, now perennially renew the face of the earth. As Mr. Proctor says : " Now

E

that astronomers have almost by unanimous consent accepted the doctrine of the development of our system which involves the belief that the whole mass of each member of the system was formerly gaseous with intensity of heat, they can no longer doubt that the moon once had seas, and an atmosphere of considerable density." [59]   The present condition of the moon Professor Newcomb makes known with one dip of his pen. "The atmosphere with which it has been covered, and the inhabitants with which it has been peopled, are no better than the products of a poetic imagination." [60]

VIII.

THE MOON'S SURFACE.

"O Moon ! the oldest shades 'mong oldest trees
Feel palpitations when thou lookest in :
O Moon ! old boughs lisp forth a holier din
The while they feel thine airy fellowship.
Thou dost bless everywhere, with silver lip
Kissing dead things to life. The sleeping kine,
Couch'd in thy brightness, dream of fields Divine :
Innumerable mountains rise, and rise,
Ambitious for the hallowing of thine eyes ;
And yet thy benediction passeth not
One obscure hiding-place, one little spot
Where pleasure may be sent : the nested wren
Has thy fair face within its tranquil ken,
And from beneath a sheltering ivy leaf
Takes glimpses of thee ; thou art a relief
To the poor, patient oyster, where it sleeps
Within its pearly house ;—the mighty deeps,
The monstrous sea is thine—the myriad sea !
O Moon ! far spooming Ocean bows to thee,
And Tellus feels her forehead's cumbrous load." [61]

THE moon presents to all observers several
aspects on several occasions. At one time she

is a luminous arc, when the poets sing of her
as the queen of the silver bow. Each succeed-
ing night she is enlarged till a semicircle is
outlined, and filled in with soft white lustre.
Soon the terminator, as it is called, or the
boundary between the shining and shaded portions, becomes more convex with brightness,
until the gibbous moon becomes full, rounded
out with radiance—one of the loveliest objects
ever seen in the dark dome of night. These
changes or phases demonstrate that the moon
is a dark body, and is illumined only by beams
borrowed from the sun. This was believed by
Thales of Miletus, one of the seven sages of
Greece; by his scholar and successor, Anaximander; by Pythagoras, Parmenides, Empedocles, and others among the ancients : while
Antiphon and others taught that her light was
her own. Some thought that her phases were
occasioned by the shadowing of the earth, which
came between. " Berosus, the Chaldean, gave
a very original explanation of the phases and
eclipses of the moon. He said it had one side
bright, and the other side just the colour of
the sky, and in turning it represented the
different colours to us. Some savage races say

that the moon when decreasing breaks up into stars, and is renewed each month by a creative act. The Indians used to say that it was full of nectar, which the gods ate up when it waned, and which grew again when it waxed." [62] An instructive Talmudic legend is to the effect that the moon at first shone with her own light; but on her becoming jealous of the sun's superior splendour, her radiance flew away and became a host of stars, leaving the dissatisfied orb to shine only with beams lent by the light of day. It was to portray her phases that "the ancients represented the moon, as they did the sun, sitting in a chariot, and drawn by horses. The old opinion was, that one of these horses was black, and the other white—a motley equipage; but we find Homer, Hesiod, and, among the Latins, Ovid, mentioning it. They intended to convey by it some idea of the moon's being sometimes dark and sometimes enlightened; but they were not so well agreed about the cattle as her chariot. Claudian, to express the rapidity of her motion, makes them stags; and we find some very old writers among the Greeks who fix upon oxen for her." [63] Some among the ancients thought the moon a large,

well-polished mirror, reflecting our own oceans
and mountains : the dark spots representing
our seas, and the bright patches our continents.
Humboldt found this fancy preserved to his
day among the people of Asia Minor. He
writes : "I was once very much astonished to
hear a very accomplished Persian of Ispahan,
who had certainly never read a Greek book,
to whom I was showing in Paris the spots on
the moon's face through a large telescope, pro-
pose the same hypothesis of reflection as that
of Agesinax, referred to in the text, as prevail-
ing in his own country. 'It is ourselves that
we see in the moon,' said the Persian, 'that is
the map of our earth.' One of the interlocutors
in Plutarch's conversation on the moon would
not have expressed himself otherwise." [64]   But,
thanks to the telescope and micrometer, which
have been at work, delineations of the moon
are now more accurate than terrestrial maps ;
every bay and promontory are seen in distinct
outline ; and photography has transferred the
likeness of the moon to metallic plates with
such fidelity, that to err in this matter is human
no longer. Still, the dissipation of such illusions
will never diminish our admiration of the moon's

fair, if not fairy, face. The dream may vanish, but the vision will remain ; for

> " A thing of beauty is a joy for ever :
> Its loveliness increases ; it will never
> Pass into nothingness."—KEATS's *Endymion.*

What a magnificent sight our world must present to lunar astronomers, with an earth-light about fourteen times the intensity of moonlight! If we admire the new moon, how they must admire the new earth ; for the earth, of course, has similar phases ! How they must adore the crescent earth and the half-earth, the gibbous earth and the full earth ! Our self-esteem forbids us to think that all this terrestrial glory is thrown away on a moon which is un-inhabited and dead. Be this as it may, " The earth is at the full at the time of the new moon, and at the new at the time of the full moon." [65] This explains why at new moon the whole of the lunar disc is dimly visible at sunset and dawn; for the earth reflects the sun's light upon the moon, just as the moon reflects it upon the earth. " The light which passes from the illuminated hemisphere of the earth to the obscure surface of the moon becomes reflected,

returns in a fainter form to the earth, and makes
visible the half of the moon, which is not only
edged with a silvery crescent, but is of a pale
and ashy tint throughout, which causes it to
stand out against the azure blue of the sky." [6]
This phenomenon is called *earth-shine.*

IX.

⌈THE MOON'S MOTIONS.

" With how sad steps, O Moon, thou climb'st the skies !
How silently, and with how wan a face !
What ! may it be, that ev'n in heavenly place
That busy archer his sharp arrows tries ?
Sure, if that long-with-love-acquainted eyes
Can judge of love, thou feel'st a lover's case ;
I read it in thy looks ; thy languished grace,
To me, that feel the like, thy state descries.
Then, e'en of fellowship, O Moon, tell me,
Is constant love deemed there but want of wit ?
Are beauties there as proud as here they be ?
Do they above love to be loved, and yet
Those lovers scorn, whom that love doth possess ?
Do they call virtue there—ungratefulness ? "

THAT the heavenly bodies move is never
disputed by us, but it was by our fathers in
the brave days of old. Nearly all of the most
ancient philosophers believed the sky to be a
solid dome or firmament, and could not admit
the idea of a star standing alone in space, or
having a free motion of its own. So still, the

North American Indians, the Creeks to wit,
believe the earth to be a plane and quiescent :
that the sun, moon, and stars wheel round the
earth ; but that some of the celestial orbs are
fixed or stuck on to the sky.[68]   Pharnaces is
reported to have feared that the moon would
fall, and pitied those who were "plumbe under
the course of the moone, lest so weightie a mass
should tumble down upon their heads."   Plutarch, who cites this odd notion, thought that
her motion would keep her up.   Pythagoras
seems to have comprehended the solar system,
so far as to have recognised the diurnal rotation
of the earth, and the revolutions of the other
planets.   But the ancient Hebrews regard the
heavens "as a canopy or a curtain, spread over
the earth in such infinite distance, that men
appear from thence 'like grasshoppers'; it is
an immeasurable tent for the habitation of God.
It is strong and massive, like a molten mirror,
but not brazen, like the Homeric heaven; it
resembles the mirror chiefly with regard to its
bright splendour, for it is like pellucid sap-
phire, or like crystal.   This vault has a gate,
through which the angels descend to the earth,
or through which the prophets beheld their

heavenly visions. It has, further, windows or doors, through which the rain and dew, snow and hail, treasured up in the clouds above, and held together in those spheres by the will of God, pour down upon the earth at His command; by which the tempests also, there confined in apartments, are let loose; and through which the lightning flashes, either as a symbol of Divine omnipotence, or as a messenger of Divine wrath. In the heaven or firmament, the sun, the moon, and the stars are fixed, to send their light to the earth and its inhabitants, and to regulate the seasons; hence the heaven is described as exercising power or government over the earth, since the phenomena of the air also are controlled by its influence. Beyond this illumined canopy reigns darkness, which the Divine wisdom has, with a nice distinction, separated from the regions of light. But above it is a sphere of liquid stores; here God dwells, for here He has framed His chambers; here is His sanctuary, His palace, the place of His glory; from hence He traverses the world on the wings of the wind and in the chariot of the clouds; for the heaven is His throne and the earth is His footstool.

That whole vault is supported by mighty pillars or foundations, resting on the earth; and thus heaven and earth are marked as one majestic edifice, forming the universe." Dr. Kalisch, the Hebrew commentator, who thus summarises the cosmography of his ancestors, justly adds, " Many of these notions, especially those concerning the abode of the Deity, are rather poetical metaphors than the real conceptions of the Hebrews ; and although some of them might be the remnants of mythic times, others are certainly figurative expressions."[69]   So much for former times. *Now* we not only know that the moon moves, but we are able to measure her motions to a second of arc or a moment of time.   Her rotation on her axis, for example, and her revolution round the earth, are both performed in 27 days, 7 hours, 43 minutes, and 11½ seconds.   This is why we see only one side of our satellite. Any person can demonstrate to himself this duplex movement in the following manner :—Let a lamp be placed in the centre of a room, to represent the earth.   Then let the observer walk half-way round the room with his face toward the lamp.   He will find that he has himself turned half-way round ; and that if he commenced with his face towards the east, it is now directed towards the west. When he regains his starting-point, he will have rotated upon his own centre while he described a circumference about the lamp. The moon's revolution round the earth is called her periodic time, or, in more sonorous phrase, her periodicity. It is also styled her sidereal movement, or her revolution in relation to the stars. Her lunar period, or lunation, is determined by the recurrence of her phases from new moon to new moon again, and is accomplished in 29 days, 12 hours, 44 minutes, and 3 seconds. This is her synodic period, or her revolution in relation to the sun. The ancient Hebrews were not far wrong in this reckoning : " The renewal of the moon comes round in not less than *twenty-nine* days and a half and forty minutes." [70] Eudoxus of Cnidus, a scholar of Plato, who flourished about 366 B.C., travelled while young into Egypt, and there, in conversation with the hierophants, learned the first lessons of regular astronomy. He made the synodic revolution of the moon to be 29 days, 12 hours, 43 minutes, 38 seconds, a difference of but 25 seconds from the calculation of to-day. The proper motion of the moon in its orbit about the earth is from east to west.

The moon's *librations*, or balancing motions, by which she appears to rock slightly to and fro, are threefold. The diurnal libration is explained by the fact that, when she rises, we view her as if from an eminence, and so see a little over her; when she has mounted to the meridian, we look her full in the face; and, at her setting, we again catch a side-face view. Her libration in latitude is thus occasioned. Her axis not being perpendicular to the plane of her orbit, but inclined $1° 32'$, her poles appear alternately for a little while to our gaze. The libration in longitude means that, as her orbital motion is not so uniform as her rotation, she sometimes exhibits more of her eastern and sometimes more of her western edge. There is a fourth perturbation, called the spheroidal, which is produced in the moon's progress by the protuberant matter of the earth's equator. This was discovered by Lagrange. But we must leave these librations.

One knotty question remains, which Mr. Norman Lockyer has untied. It concerns the moon's *nodes.* " The plane in which the moon performs her journey round the earth is inclined 5° to the plane of the ecliptic, or the plane in which the earth performs her journey round the sun. The two points in which the moon's orbit, or the orbit of any other celestial body, intersects the earth's orbit, are called the nodes. The line joining these two points is called the line of nodes. The node at which the body passes to the north of the ecliptic is called the ascending node, and the other the descending node." [71] In passing, we may note here a point of analogy between our astronomy and that of the Brahmins. " In the language of their rules we may trace some marks of a fabulous and ignorant age, from which indeed even the astronomy of Europe is not altogether free. The place of the moon's ascending node is with them *the place of the dragon,* or *the serpent ;* the moon's distance from the node is literally translated by M. Legentil *la lune offensée du dragon.* Whether it be that we have borrowed these absurdities from India, along with astrology, or that the popular theory of eclipses has, at first, been everywhere the same, the moon's node is also known with us by the name of the *cauda draconis* " [72] (the dragon's tail).

An important use of the heavenly bodies was made ages before their motions were either measured or marked. They served men as chronometers, " for signs and for seasons, and for days and years." Time is defined by Dr. Johnson as " the measure of duration," and he illustrates his definition with a quotation from Locke's *Human Understanding.* But we think that a better definition of time is, " the measure of motion." Locke argues against this ; but Sir William Hamilton, Dr. Thomas Brown, Kant, and others, maintain that time involves succession, which is a series of divisible motions. The earth moves, the sun, moon, and stars move, nothing stands still : we divide this progress into parts called years, months, days ; and the sum total we denominate time. We now employ mechanical contrivances, of wonderful workmanship, to subdivide our days into hours and minutes and seconds. But the ancients cared little for minute fragments ; their time-piece by day was the shine and shadow of the sun, and by night the progress and phases of the moon. The etymology of the word is full of meaning. *Moon* and *month* are twins, whose parentage was Sanskrit. Professor Max Müller says, " The moon also was called *mâs*, the measurer, which is its actual name in Sanskrit, closely connected with Greek μείς, Latin *mensis*, English *moon.*" [73] In Egyptian also Techu is the name of the moon-god, and is called " the measurer of this earth, the distributor of time." [74] The Persian *mâh*, for moon, has the same idea at its root. The German *mond* and *monat* signify moon and month. Although it is doubtful whether the early Hebrews did anything in lunar computation, it is certain that the word *chodesh* (חֹדֶשׁ), which means a month, is derived from *châdash* (חָדַשׁ), *to renew*, and so expresses a month beginning with the new moon. The Hebrews designated their months in numerical order, the first, second, third, and so on. Geminus of Rhodes says that " the system pursued by the ancient Greeks was to determine their months by the moon, and their years by the sun." The Athenians began their year upon the first new moon after the summer solstice. This year they divided into twelve months, containing alternately 30 and 29 days. Each month was

F

divided into three decades of days, the first day
being called νεομηνία, because it fell on the new
moon.[75]   Among the Romans, Romulus is said
to have divided the year into ten months ;
but Numa, in imitation of the Greeks, increased
the number to twelve, according to the course
of the moon.   But Julius Cæsar adjusted the
year to the course of the sun, assigning to each
month the number of days which it still contains.   The Romans also divided their months
into three parts.   The first day was called
Calendæ, from an old verb meaning to *call out*,
because a pontiff then made proclamation to
the people that it was new moon.[76]   These
calendæ have given us our word calendar.   Dr.
Inman tells us that, " amongst the Indians of
North America, time was computed by months
or moons, and ' beaver moon,' ' buck moon,'
' buffalo moon,' and the like formed their sole
calendar.   It was the same amongst the early
Greeks, who had their ' planting moons,' ' reap-
ing moons,' ' wine moons,' and the like.   A
similar plan was adopted by the French, when
they revolutionised almost everything which
had previously been honoured by Church and
State."[77]   Dr. Pickering found Arabised ne-

grues in Zanzibar who availed themselves of
the same lunar chronometry. "The Soahili,
besides the usual Muslim calendar, have one of
their own. Their new year commenced, in
1844, on the 29th of August, or, more precisely,
at 6 p.m. on the evening of the 28th; and I
remarked further that it immediately followed
full moon. Sadik stated that the Soahili year
'consists of twelve moons and ten days; and
that from the weather on these supernumerary
days the people prognosticate that of the whole
year. The months or moons are numbered,
and three only have names, Shaban (understood
to indicate the time of planting), Rejëb, and
Ramadan,' appellations which are well known
in the Muslim calendar. Indeed it was reiter-
ated 'that the Soahili year is the same with the
Arab, and consists in like manner of three hundred and sixty-five days, or of twelve moons
and ten days,' a statement which seems to refer
to some agricultural calendar used in Southern
Arabia." In like manner the Taheitians "measure long periods of time by moons, or luna-
tions. They appeared to have no measure-
ments for short distances or short periods of
time, corresponding to a mile or an hour, but

always pointed to the place in the heavens where the sun would be when we should arrive at the proposed station." [78]   The North American Indians still use this simple system. Schoolcraft says, " The new year commences with the Creeks immediately after the celebration of the busk, at the ripening of the new corn in August.   They divide the year into two seasons only, to wit, winter and summer, and subdivide it by the successive moons, beginning the winter with the moon of August, called the big ripening moon.   September, little chestnut moon ; October, big chestnut moon ; November, falling leaf moon ; December, big winter moon; January, little winter moon, *alias* big winter moon's young brother; February, the windy moon; March, little spring moon ; April, big spring moon; May, mulberry moon ; June, blackberry moon ; July, little ripening moon." [79]   From the same author we learn that the Comanches, " for short periods, past or future, count by moons, from full to full." [80] The Winnebagoes reckon twelve moons for a year, of which spring is the commencement. " They differ somewhat in the names of their twelve moons.   The following, however, is the

common almanack among them :—1st moon
Drying the earth; 2nd, Digging the ground,
or planting corn; 3rd, Hoeing corn; 4th, Corn
tasselling; 5th, Corn popping, or harvest time;
6th, Elk whistling; 7th, Deer running; 8th,
Deer's horns dripping; 9th, Little bear's time;
10th, Big bear's time; 11th, Coon running;
12th, Fish running." " The moon is not considered by them as having influence on men,
vegetation, or animals, and no regard is paid
to the particular time of the moon's phases in
planting corn and other seed." [81]   The Kenis-
tenos "divide the year by the succession of
moons. The names which they give to the
moons are descriptive of the several seasons.
They are in their order, beginning with the
month of May, called the frog moon; the moon
when birds begin to lay their eggs; the moon
when birds moult, or cast their feathers; the
moon when birds begin to fly; the moon in
which the moose casts its horns; the ratting
moon; hoar-frost moon, or ice moon; whirl-
wind moon; cold moon; big moon; eagle moon;
and goose moon, which is their April." [82]   So
the poor Indian's mind is not altogether "un-
tutored "; his horometer is in the heavens, and
his dial-plate is the moon.

But if the valuable mechanical science of horology has rendered us somewhat independent in civil life of the celestial orbs, it is pleasing to find that the moon still affords good aid in the country districts of our own land. " The parish lantern " is a common expression for the lunar lamp in Cornwall, Berkshire, Worcestershire, Yorkshire, and the midland counties, and indeed all over England. Sir Walter Scott tells us that the moon was called " Mac Farlane's lantern " in Scotland : for " the clan of Mac Farlane, occupying the fastnesses of the western side of Loch Lomond, were great depredators on the low country; and as their excursions were made usually by night, the moon was proverbially called their lantern. Their celebrated pibroch of *Hoggil nam Bo,* which is the name of their gathering tune, intimates similar practices, the sense being,—

> " We are bound to drive the bullocks,
>   All by hollows, hirsts, and hillocks,
>   Through the sleet and through the rain.
>   When the moon is beaming low
>   On frozen lake and hills of snow,
>   Bold and heartily we go ;
>   And all for little gain." [83]

Whilst we are writing upon the moon's

motions, we may dispel a delusion which is
still current. It is a very prevalent opinion in
many parts that the harvest moon invariably
occurs at the time of harvest, whenever that
may take place. This notion is thus referred
to and refuted by an American author : "About
the time of the autumnal equinox, the moon,
when near her full, rises about sunset a number
of nights in succession. This occasions a remarkable number of brilliant moonlight evenings ; and as this is, in England, the period of
harvest, the phenomenon is called the *harvest
moon.* Its return is celebrated, particularly
among the peasantry, by festive dances, and
kept as a festival, called the *harvest home,*—an
occasion often alluded to by the British poets.
Thus Henry Kirke White :—

> ' Moon of harvest, herald mild
> Of plenty, rustic labour's child,
> Hail, O hail ! I greet thy beam,
> As soft it trembles o'er the stream,
> And gilds the straw-thatched hamlet wide,
> Where innocence and peace reside ;
> 'Tis thou that glad'st with joy the rustic throng,
> Promptest the tripping dance, th' exhilarating song.'

"To understand the reason of the harvest
moon, we will, as before, consider the moon's

orbit as coinciding with the ecliptic, because we may then take the ecliptic, as it is drawn on the artificial globe, to represent that orbit. We will also bear in mind that, since the ecliptic cuts the meridian obliquely, while all the circles of diurnal revolution cut it perpendicularly, different portions of the ecliptic will cut the horizon at different angles. Thus, when the equinoxes are in the horizon, the ecliptic makes a very small angle with the horizon ; whereas, when the solstitial points are in the horizon, the same angle is far greater. In the former case, a body moving eastward in the ecliptic, and being at the eastern horizon at sunset, would descend but a little way below the horizon in moving over many degrees of the ecliptic. Now, this is just the case of the moon at the time of the harvest home, about the time of the autumnal equinox. The sun being then in Libra, and the moon, when full, being, of course, opposite to the sun, or in Aries, and moving eastward, in or near the ecliptic, at the rate of about thirteen degrees per day, would descend but a small distance below the horizon for five or six days in succession—that is, for two or three days before, and the same number of days after, the full—and would consequently rise during all these evenings nearly at the same time— namely, a little before, or a little after, sun- set—so as to afford a remarkable succession of fine moonlight evenings." [84]

Something may be here appropriately said with regard to the supposed arrest of the moon's motion at the command of Joshua. We do not for a moment hesitate to pro- nounce the story to be apocryphal in essence and poetical in form. It never should have been taught as miracle or history. Three ob- jections are fatal to its acceptance as fact. First, it is *unscientific.* Whatever those who are addicted to the dogma of Biblical infalli- bility may now advance in their attempt to reconcile Scripture with science, the almost unanimous opinion hitherto held of the occur- rence has been that the sun and moon actually stood still during a whole day. Dr. Richard Watson's fourth letter of his *Apology for the Bible*, in reply to Thomas Paine, is said in Smith's *Dictionary of the Bible* to have forcibly stated the "literal and natural interpretation of the text as intended to describe a miracle."

This was the "orthodox" view. Theories of an optical illusion, and so forth, are but forced and feeble efforts to evade the catholic interpretation, till recently accepted as true, though now resigned as truthless. Honesty is the best policy in scriptural hermeneutics as in everything else; and it is not honest to maintain a popular tenet as a matter of life or death till it is shown to be untenable, and then to veer round with the wind merely to save appearances. It is better to acknowledge frankly that our Biblical boundaries are held subject to the rectifications of frontier which science may require. So shall we save ourselves from the stultification which results from commitment to an exegesis in one age which has to be abandoned in the next. Scripture writers and readers may err, for they are human; nature cannot, for it is Divine. Now science demonstrates that for the sun to stop in his course means that the earth must stand still, for in relation to the earth the sun is stationary. But Dr. Thomson "computed that if by any sudden shock the earth were arrested in its orbit, the heat generated by the impulse would be equal to 11·200 degrees of the centigrade thermometer, even if the capacity of our planet for heat were as low as that of water; it would therefore be mostly reduced to vapour, and should the earth then fall to the sun, as it certainly would do, the quantity of heat developed by striking on the sun would be 400 times greater." [85] But no physical catastrophe took place in the days of Joshua, *ergo*, the miracle is a myth. Secondly, the story is *unmoral*. Such a stupendous sign, had it happened, would have had a worthy purpose. What was the end to be served on this occasion? The mightiest miracle on record was wrought to enable a bloodthirsty people to "avenge themselves upon their enemies"! *Credat Judæus Apella*. It is a wicked libel upon the God of love to imagine that He would interrupt the harmony of the solar system for the sake of a "great slaughter." It would be a devil's deed; and therefore, in the name of the Father of all, Amorites as well as Israelites, we declare the legend false. Thirdly, the story is *unhistorical*. That is to say, Joshua never commanded the sun and moon to stand still; or, if he uttered the bold apostrophe, he knew nothing of the existing record. The unknown author or editor of the Book of Joshua foisted the fiction in as a quotation from the Book of Jasher, which could not have been written before the time of David (see 2 Sam. ii. 18), 500 years after Joshua, and which was but a collection of floating traditions and fragmentary songs, without historic foundation. The *Speaker's Commentary* tears up the tale by the roots : " Is the Book of Joshua committed to the upholding of this marvel in the heavens as having actually taken place? Answer may perhaps reasonably be given in the negative. The whole passage may, and even ought, on critical grounds, to be taken as a fragment, of unknown date and uncertain authorship, interpolated into the text of the narrative, the continuity of which is broken by the intrusion " (*Joshua*, p. 58). We shall best conserve what is Divine in the Bible by conceding what is human therein ; while the deification of error threatens the dethronement of truth.

Before we quit the moon's motions, a word may be in season respecting the myth of the music of the spheres, which has come down to us from the school of Pythagoras, the first

" philosopher," or wisdom-lover, all of whose
thoughts upon the mystery of number, as
Maurice says, combined themselves with musical feelings and associations. " Whence came
that strange disposition of thoughts and words
into verse ? Whence the fascination of melody
and tune ? Whence, if number be not the
secret law, the moving soul of the universe ? " [86]
Pythagoras and his followers seem to have
ascribed to the stars a certain musical rhythm ;
they were supposed to create harmony in their
orbits, and thus to keep the motions of the
universe in time and tune. Adam Smith,
alluding to these " proportionable motions,"
adds, " It seems to have been the beauty of
this system that gave Plato the notion of something like an harmonic proportion to be discovered in the motions and distances of the
heavenly bodies, and which suggested to the
earlier Pythagoreans the celebrated fancy of the
music of the spheres ; a wild and romantic idea,
yet such as does not ill correspond with that
admiration which so beautiful a system, recom-
mended too by the graces of novelty, is apt to
inspire." [87] Our encyclopedian Shakespeare, of
course, alludes to this celestial concert :—

" There's not the smallest orb which thou behold'st,
    But in his motion like an angel sings,
    Still quiring to the young-eyed cherubim." [88]

And Jean Baptiste Rousseau, with all of a
Frenchman's fervour, sings :—

" Les cieux instruisent la terre
    A révérer leur Auteur.
    Tout ce que leur globe enserre
    Célèbre un Dieu créateur.
    Quel plus sublime cantique
    Que ce concert magnifique
    De tous les célestes corps !
    Quelle grandeur infinie !
    Quelle divine harmonie
    Résulte de leurs accords ! " [89]

The music of the spheres was a dream, no
doubt; and, like other visions of childhood,
it is doomed to vanish with the manhood of
the world.   But we can never forget these
fancies altogether ; for which of us would be
now a man if he had not been first a child ?
But if the wheels of nature turn silently on
their axles, if the burning seraphim which glow
through infinite space have no voice of their
own, they lack no chorus to accompany their
sublime movements.   There is a " music of the
moon," which " sleeps in the plain eggs of the

nightingale," and awakes to "the beauty of a thousand stars." There is an early bird that calls the sun, when

> "at heaven's gate she claps her wings,
> The morn not waking till she sings."

And above all the feathered tribes is heard the swell of the saintly Psalm, whose oriental beauty and Divine melody no science will ever be able or willing to mar :—

> " The heavens are declaring the glory of God,
> And the expanse is displaying the work of His hands.
> Day unto day is uttering speech,
> And night unto night is breathing out knowledge.
> There is no speech and there are no words ;
> Their voice is not to be heard."

> " What though, in solemn silence, all
> Move round the dark terrestrial ball ;
> What though no real voice nor sound
> Amidst their radiant orbs be found ;
> In reason's ear they all rejoice,
> And utter forth a glorious voice ;
> For ever singing as they shine,
> 'The Hand that made us is Divine !'"

NOTES.

1. The Poetical Works of William Broome; Cooke's Edition. London, 1796, p. 81.
2. *Meditations and Contemplations*, by James Hervey, A.M. London, 1824, p. 234.
3. *Historical Survey of the Astronomy of the Ancients*, by Sir G. C. Lewis. London, 1862, p. 2.
4. *Critique of Practical Reason*, by Immanuel Kant. London, 1879, p. 376.
5. *Astronomy*, by J. Rambosson. Translated by C. B. Pitman. London, 1875, p. 381.
6. *The Moon*, by Edmund Neison. London, 1876, p. 85.
7. *Outlines of Astronomy*, by Sir John F. W. Herschel, Bart. London, 1869, p. 282.
8. *Mappa Selenographica.* Auctoribus Guil. Beer et Joanne H. Maedler. Berolini, 1834.
9. *Mondcharte*, von W. G. Lohrmann. Herausgegeben von Dr. J. F. J. Schmidt. Leipzig, 1878.
10. *Dar-thula. The Poems of Ossian.* Translated by James Macpherson. London, 1807.
11. *Elementary Lessons in Astronomy*, by J. Norman Lockyer, F.R.S. London, 1876, p. 89.
12. *Popular Astronomy*, by George Biddell Airy, Astronomer Royal. London, 1866, p. 67.

G

13. *God's Glory in the Heavens,* by William Leitch, D.D. London, 1863, p. 27.
14. *Popular Astronomy,* by François Arago. London, 1858, vol. ii. p. 416.
15. *A Manual of Budhism,* by R. Spence Hardy. London, 1880. Note to p. 23.
16. *Astronomical Myths,* based on Flammarion's *History oj the Heavens.* London, 1877, p. 172.
17. *Dissertation on the Knowledge of the Ancients in Astronomy,* by John Rogers, M.D. London, 1755, p. 120
18. *The Spectator, No.* 565, by Joseph Addison.
19. *The Poetical Works of Lord Byron :* Childe Harold's · Pilgrimage, iv. 27.
20. *The Treasury of Science,* by Frederick Schoedler, Ph.D. London, 1865, p. 166.
21. *The Midnight Sky,* by Edwin Dunkin, F.R.S. London, 1879, p. 263.
22. *Descriptive Astronomy,* by George F. Chambers, F.R.A.S. Oxford, 1867, p. 81.
23. *A Treatise on Astronomy,* by Elias Loomis, LL.D. New York, 1870, p. 118.
24. *Popular Astronomy,* by Simon Newcomb, LL.D. New York, 1882, p. 323.
25. *A Complete System of Astronomy,* by the Rev. S. Vince, M.A., F.R.S. London, 1814, ii. p. 255.
26. *The Childhood of Religions,* by Edward Clodd, F.R.A.S. London, 1875, p. 11.
27. *Heroes of Science :* Astronomers, by E. J. C. Morton. London, 1882, p. 13.
28. *The Talmud,* by Joseph Barclay, LL.D. London, 1878, p. 150.
29. Hardy's *Manual of Budhism,* p. 20.
30. *Ibid.,* p. 11.

31. *A Dictionary in Sanskrit and English*, by H. H. Wilson. Calcutta, 1874, p. 712.
32. *The Marvels of the Heavens*, by Camille Flammarion. London, 1870, p. 256.
33. *The Connexion of the Physical Sciences*, by Mary Somerville. London, 1877, p. 70.
34. *The Poetical Works of James Thomson:* Autumn, 1088–1102. London, 1850.
35. *A History of New York*, by Diedrich Knickerbocker. London, 1839, p. 1.
36. Richard A. Proctor, in the *Cornhill Magazine* for August, 1873, p. 179.
37. Lewis's *Astronomy of the Ancients*, p. 3.
38. *The Poetical Works of Thomas Campbell :* Moonlight. London, 1858, p. 414.
39. *The History of Greece*, by George Grote, D.C.L London, 1872, i. 317.
40. Hardy's *Manual of Budhism*, p. 20.
41. *The Poetical Works of Samuel Butler* (Gilfillan's Edit.). Edinburgh, 1854, i. 176.
42. *The Works of Rabelais.* London, 1871, p. 27.
43. Grosart's Fuller's Worthies Library. *Jacke Jugeler* (assumed date, 1563), vol. iv. p. 526.
44. *Popular Tales of the West Highlands*, orally collected by J. F. Campbell. Edinburgh, 1860, i. 272.
45. *Selections from the Talmud.* Translated by H. Polano. London, p. 216.
46. *Vestiges of the Natural History of Creation*, by Robert Chambers. London, 1860, p. 20.
47. *The Poetical Works of John Milton :* Paradise Lost, Book iv.
48. *The Moon*, by J. Nasmyth and J. Carpenter. London, 1874, p. 18.

49. *The Cornhill Magazine* for August, 1873, p. 186.
50. *Miscellaneous Essays,* by H. T. Colebrooke. London, 1873, i. 171.
51. *The History of Mexico,* by Abbé D. Francesco Saverio Clavigero. Translated from the Italian by Charles Cullen. London, 1787, i. 248.
52. *The Childhood of the World,* by Edward Clodd. London, 1875, p. 62.
53. *The Works of Ben Jonson:* The Song of Hesperus. From *Cynthia's Revels.*
54. *Elements of Physiophilosophy,* by Lorenz O'Kew, M.D. London, 1847, pp. 52, 56.
55. *The Poetical Works of Percy Bysshe Shelley.* London (Moxon), p. 543.
56. *The Marvels of the Heavens,* by Camille Flammarion. London, 1870, p. 247.
57. *A New Theory of the Earth,* by William Whiston, M.A. Cambridge, 1708, p. 368.
58. *The Heavens,* by Amédée Guillemin. Fourth Edition. London, 1871, p. 143.
59. *The Poetry of Astronomy,* by R. A. Proctor. London, 1881.
60. *Popular Astronomy,* by Simon Newcomb, LL.D. New York, 1882, p. 315.
61. *The Poetical Works of John Keats:* Endymion, Book III. London, 1876.
62. *Astronomical Myths,* based on Flammarion, by John F. Blake. London, 1877, p. 220.
63. *Urania: or a Compleat View of the Heavens,* by John Hill, M.D. London, 1754. Art., "Moon."
64. *Cosmos,* by Alexander von Humboldt. Sabine's Edition, p. cxxx.
65 *A Complete System of Astronomy,* by the Rev. S. Vince, M.A., F.R.S. London, 1814, i. 207.

66. *Astronomy*, by J. Rambosson. London, 1875, p. 201.

67. *Sonnets*, by Sir Philip Sidney.

68. *Historical and Statistical Information of the Indian Tribes*, by Henry R. Schoolcraft. Philadelphia, i. 271.

69. *Historical and Critical Commentary :* Genesis, by M. M. Kalisch, Ph.D., M.A. London, 1858, p. 14.

70. *A Talmudic Miscellany.* London, 1878, p. 181.

71. *Elementary Lessons in Astronomy*, by J. Norman Lockyer, F.R.S. London, 1876, p. 90.

72. *Remarks on the Astronomy of the Brahmins.* Works of John Playfair. Edinburgh, 1822, p. 150.

73. *Hibbert Lectures on the Religions of India*, by F. Max Müller, M.A. London, 1878, p. 187.

74. *Hibbert Lectures on the Religion of Ancient Egypt*, by P. Le Page Renouf. London, 1880, p. 116.

75. *The Antiquities of Greece*, by John Potter, D.D. Edinburgh, 1818, p. 524.

76. *Roman Antiquities*, by Alexander Adam, LL.D. London, 1825, pp. 300–302.

77. *Ancient Faiths embodied in Ancient Names*, by Thomas Inman, M.D. London, 1873, ii. p. 858.

78. *The Races of Man*, by Charles Pickering, M.D. London, 1851, pp. 194, 67, 63.

79. *Information respecting the History, Condition, and Prospects of the Indian Tribes of the United States*, by H. R. Schoolcraft, LL.D. Philadelphia, Part V., p. 276.

80. *Ibid.* Part I., p. 237.

81. *Ibid.* Part IV., pp. 239, 240.

82. *Ibid.* Part V., p. 171.

83. *Waverley ; or, 'Tis Sixty Years Since*, by Sir Walter Scott, Bart. Note to chap. xxxviii.

84. *Letters on Astronomy*, by Denison Olmsted, A.M. Boston (U. S.), 1842, p. 177.

85. *On Molecular and Microscopic Science*, by Mary Somerville. London, 1869, i. 27.

86. *Moral and Metaphysical Philosophy.* Part I. : Ancient Philosophy, by F. D. Maurice, M.A. London, 1850, p. 95.

87. *Essays*, by Adam Smith, LL.D., F.R.S. London, 1869, p. 345.

88. *The Merchant of Venice*, Act V., Scene 1.

89. *Œuvres Complètes de J. B. Rousseau.* Tome Premier. Paris, 1797, p. 3.

INDEX OF NAMES, Etc.

Playfair, John.
Plutarch.
Potter, Dr. John.
Proctor, R. A.
Rabbi Meir.
Rabelais.
Rambosson, J.
Renouf, Page.
Rogers, Dr. J.
Romans.
Rousseau, J. B.
Rutherford, L. M.
Schmidt, Julius.
Schoedler, F.
Schoolcraft, H. R.
Scott, Sir W.
Shakespeare.
Shelley.
Sidney, Sir P.
Smith, Adam.

Socrates.
Somerville, Mrs.
Southey.
Stoics.
Tacitus.
Talmud.
Tennyson.
Thales.
Thomson, James.
Thomson, Sir Wm.
Tyndall, John.
Verne, Jules.
Vince, S.
Whiston, W.
White, H. K.
Wilson, H. H.
Wilson, John.
Winnebagoes.
Xenophanes.
Zöllner.

Butler & Tanner, The Selwood Printing Works, Frome, and London.

SCIENTIFIC BOOKS.

CRUISE OF H. M. S. "ALERT." Four Years in Patagonian, Polynesian, and Mascarene Waters. By R. W. COPPINGER, M.D. (Staff-surgeon on board). With 16 Plates, and several Cuts in the text, from drawings and photos by the Author and F. NORTH, R.N. Fourth Edition. 8vo, cloth, gilt edges, 6s.

"No one could be better fitted than Dr. Coppinger to put into a readable shape the result of his investigations as a naturalist, and his impressions of strange scenery and savage men. . . . Every page attests his method and his practical familiarity, etc."—*Saturday Review.*

THE EVOLUTION OF FLOWERS. By GRANT ALLEN.   [*In preparation.*

ALPINE PLANTS. Painted from Nature, by J. SEBOTH, with descriptive text by A. W. BENNETT, M.A., B. Sc. 4 vols., each with 100 Coloured Plates. Super-royal 16mo, half persian, gilt tops, each 25s.

TOURISTS' GUIDE TO THE FLORA OF THE ALPS. Edited from the work of PROF. K. W. V. DALLA TORRE, and issued under the auspices of the German and Austrian Alpine Club in Vienna. By A. W. BENNETT, M.A., B.Sc. Elegantly printed on very thin but opaque paper, 392 pp., bound as a Morocco pocket book, pocket size, 5s.

"Excellent for its purpose, and reflects credit on all concerned."—*Athenæum.*

FLOWERS AND FLOWER-LORE. By Rev.
HILDERIC FRIEND, F.L.S. Illustrated. One Vol., 7s. 6d.

CONTENTS.—The Fairy Garland—From Pixy to Puck—The Virgin's Bower—Bridal Wreaths and Bouquets—Garlands for Heroes and Saints—Traditions about Flowers—Proverbs of Flowers—The Seasons—The Magic Wand—Superstitions about Flowers—Curious Beliefs of Herbalists—Sprigs and Sprays in Heraldry—Plant Names—Language of Flowers—Rustic Flower Names—Peculiar Usages—Witches and their Flower-lore.

"We are practising real self-denial in giving only a short notice to 'Flower-lore.' . . . . It introduces us to a whole library of plant-lore ; indeed, throughout, the book is as painstaking as it is interesting."—*Graphic.*

"So thorough and so interesting, and at the same time so simple and poetic."—*Pall Mall Gazette.*

THE WANDERINGS OF PLANTS AND
ANIMALS. By PROF. VICTOR HEHN. Edited by J. STEVEN STALLYBRASS. Demy 8vo, cloth, 16s.

"No more interesting work can be imagined. . . . A profusion of learning is spent on every chapter ; at every turn some odd piece of classical lore turns up. Every student of nature, as well as every scholar, will be grateful to Mr. Stallybrass for his book. He gives them in their own tongue a great body of erudition and a collection of striking facts. The index is excellent, and particular attention should be drawn to the notes, which are most valuable, and run to one hundred pages."—*Academy* (Rev. M. G. Watkins).

"It is impossible here to give any idea of the extreme wealth of illustration. . . . It is a storehouse of entertainment. . . . Prof. Hehn writes like a living man, and not as a Dryasdust, and many of our readers will find his work supremely interesting."—*Field.*

BIRDS AND THEIR WAYS. By J. ARMSTRONG.
Illustrated. Crown 8vo, cloth gilt, gilt edges, 1s. 6d.

"Written with much ease and grace, and in an interesting manner. . . . Will form an acceptable gift-book for children of all ages."—*Practical Teacher.*

THE DYNAMO: How Made and How Used.
By S. K. BOTTONE. Numerous Cuts. Crown 8vo, cloth, 2s. 6d.

"Exceedingly plain, clear instructions for the manufacture of small dynamos."—*Journal of Science.*

POND LIFE: Insects. By E. A. BUTLER. Crown 8vo,
1s.

"Is admirably suited to convey a marvellous amount of erudition in a very few words."—*Whitehall Review.*

THE INSECT HUNTER'S COMPANION. By
REV. J. GREENE. Third Edition. Cuts. 12mo, boards, 1s.

TABULAR VIEW OF GEOLOGICAL SYS-
TEMS. By DR. E. CLEMENT. Crown 8vo, limp cloth, 1s.

HANDBOOK OF ENTOMOLOGY. By W. F. KIRBY
(Brit. Mus.). Illustrated with several hundred figures. 8vo, cloth gilt, 15s.

"It is, in fact, a succinct encyclopædia of the subject. Plain and perspicuous in language, and profusely illustrated, the insect must be a rare one indeed whose genus—and perhaps even whose species—the reader fails to determine without difficulty. . . . The woodcuts are so admirable as almost to cheat the eye, familiar with the objects presented, into the belief that it is gazing upon the colours which it knows so well. . . . Advanced entomologists will obtain Mr. Kirby's fine volume as a handy book of reference; the student will buy it as an excellent introduction to the science and as an absolutely trustworthy text-book."—*Knowledge.*

EVOLUTION AND NATURAL THEOLOGY.
By W. F. KIRBY (Brit. Mus.). Crown 8vo, cloth, 4s. 6d.

"A book of much interest from the pen of a ready writer."—*Knowledge.*

THE YOUNG COLLECTOR'S HANDBOOK OF BUTTERFLIES, MOTHS, AND BEETLES. By W. F. KIRBY (Brit. Mus.). Crown 8vo, cloth, 1s.

"A really admirable and absurdly cheap manual. The incipient entomologist will do himself an injustice if he does not procure it. Not the least striking thing in it is the faithful way in which insect markings are reproduced in the mere black and white of wood-engraving."—*Knowledge.*
"The author is an entomologist of repute. . . . His book conveys a great deal of information."—*Times.*
"Excellently proportioned."—*Saturday Review.*

THE YOUNG COLLECTOR'S HANDBOOK OF MOSSES. By J. E. BAGNALL. Numerous Woodcuts. Crown 8vo, cloth, 1s.

"Really a wonderful shilling's worth. It is an excellent introduction to the study of mosses. Any one who knows Mr. Bagnall would naturally expect such a guide. . . . No one can hesitate to order copies."—*Grevillea.*
"The illustrations are numerous and good. . . . A capital little book."—*Athenæum.*
"We do not think any botanist could have been better selected than Mr. Bagnall."—*Science Gossip.*

THE YOUNG COLLECTOR'S HANDBOOK OF SEAWEEDS, SHELLS, AND FOSSILS. By PETER GRAY and B. B. WOODWARD. With Cuts. Crown 8vo, cloth, 1s.

"We must give credit for excellent intentions on the part of both authors and publishers."—*Athenæum.*

THE YOUNG COLLECTOR'S HANDBOOK OF ENGLISH COINS AND TOKENS. With a Chapter on Greek Coins. By BARCLAY V. HEAD (Brit. Mus.). Illustrated. Crown 8vo, cloth, 1s.

"Those who take an interest in numismatics will find this volume a valuable aid to study."—*Newcastle Weekly Chronicle.*

LIFE HISTORIES OF PLANTS. By PROF. D.
Mc ALPINE. With an Introduction to the Comparative Study of Plants and Animals on a Physiological Basis. Illustrated. Royal 16mo.

HANDBOOK OF THE DISEASES OF PLANTS.
By PROF. D. MC ALPINE. Illustrated. Demy 8vo. [*In preparation.*

SCIENTIFIC ROMANCES. By C. H. HINTON.

Crown 8vo gilt, gilt top, 6s.; or in 5 parts, 1s. each.

GHOSTS EXPLAINED.

1. WHAT IS THE FOURTH DIMENSION ?

"A short treatise of admirable clearness. . . . Mr. Hinton brings us, panting but delighted, to at least a momentary faith in the Fourth Dimension, and upon the eye of this faith there opens a vista of interesting problems. . . . His pamphlet exhibits a boldness of speculation, and a power of conceiving and expressing even the inconceivable which rouses one's faculties like a tonic."—*Pall Mall.*

THE MYSTERY OF PLEASURE AND PAIN.

2. THE PERSIAN KING ; or, The Law of the Valley.

" A very suggestive and well-written speculation, by the inheritor of an honoured name."—*Mind.*
" Will arrest the attention of the reader at once."—*Knowledge.*

3. CASTING OUT THE SELF.

4. A PLANE WORLD.

5. A PICTURE OF OUR UNIVERSE.

DICTIONARY OF BRITISH BIRDS. By COLONEL
MONTAGUE. New Edition. Edited by E. NEWMAN, F.L.S. Demy 8vo, cloth gilt, 7s. 6d.

THE MICROSCOPE: Theory and Practice. By
PROF. C. NAEGELI and PROF. S. SCHWENDENER. With about 300 Woodcuts. Demy 8vo, cloth, 21s. [*In the press.*

HISTORY OF BRITISH FERNS. By E. NEWMAN,
F.L.S. Third Edition. Cuts. Demy 8vo, cloth, 18s.

A "PEOPLE'S EDITION" OF THE ABOVE

(Abridged), containing numerous figures. Fifth Edition. 12mo, cloth, 2s.

BIBLIOGRAPHY: Index and Guide to Climate.

By A. RAMSAY, F.G.S. Cuts. Demy 8vo, cloth gilt, 16s.

A SEASON AMONG THE WILD FLOWERS.

By REV. H. WOOD. Second Edition. Cuts. Crown 8vo, cloth gilt, gilt edges, 3s. 6d.

"Perfectly free from misplaced raptures, the book is also attractive from its correctness. The plates are unusually, and, indeed, remarkably good for a cheap and popular treatise."—*Academy.*

THE NATURALIST'S DIARY: A Day-Book of Meteorology, Phenology, and Rural Biology.

By CHARLES ROBERTS, F.R.C.S., L.R.C.P., etc. Crown 8vo, limp cloth, 2s. 6d.

"The 'Naturalist's Diary' is intended to be used as a work of reference on many questions relative to Natural History, Climate, Periodic Phenomena, and Rural Economy; and as a Journal in which to record new facts and observations of a similar kind. Each page is divided into two columns, the left-hand containing the printed information, notes as to what to observe or expect, etc., and the right being left for MS. additions, notes, etc.

"A delightful device. . . . Will make every man his own White of Selborne."—*Saturday Review.*

ELEMENTARY TEXT-BOOK OF BOTANY.

By PROF. W. PRANT and S. W. VINES, D.Sc., M.A., Fellow and Lecturer of Christ's College, Cambridge. Fourth Edition (1885). Illustrated by 275 woodcuts. Demy 8vo, cloth, 9s.

ELEMENTARY TEXT-BOOK OF ZOOLOGY.

By PROF. W. CLAUS (University Vienna), and PROF. A. SEDWICK (Trin. Coll., Camb.). With 706 new woodcuts. Demy 8vo. SECTION I., General Introduction and Protozoa to Insecta, 21s. SECTION II., Mollusca to Man, 16s.

"Teachers and students alike have been anxiously waiting for its appearance. . . . We would lay especial weight on the illustrations of this work for two reasons; firstly, because correct figures are of enormous assistance to the student, . . . and secondly . . . it contains as rich a supply of well-drawn, well-engraved, and well-selected figures as ever man could desire. . . . Are admirably printed. . . . The whole enterprise reflects the greatest credit."—*Zoologist.*

www.ingramcontent.com/pod-product-compliance
Lightning Source LLC
Chambersburg PA
CBHW021946190326
41519CB00009B/1161